Connected Mathematics™

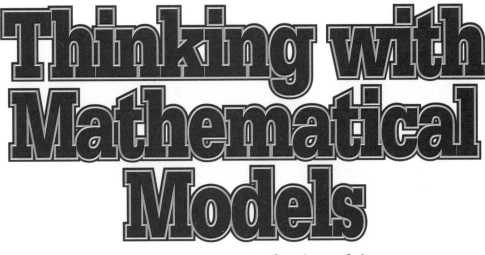

Thinking with Mathematical Models

Representing Relationships

Student Edition

Glenda Lappan
James T. Fey
William M. Fitzgerald
Susan N. Friel
Elizabeth Difanis Phillips

Developed at Michigan State University

DALE SEYMOUR PUBLICATIONS®

Connected Mathematics™ was developed at Michigan State University with the support of National Science Foundation Grant No. MDR 9150217.

This project was supported, in part,
by the
National Science Foundation
Opinions expressed are those of the authors
and not necessarily those of the Foundation

The Michigan State University authors and administration have agreed that all MSU royalties arising from this publication will be devoted to purposes supported by the Department of Mathematics and the MSU Mathematics Education Enrichment Fund.

This book is published by Dale Seymour Publications®, an imprint of Addison Wesley Longman, Inc.

Managing Editor: Catherine Anderson
Project Editor: Stacey Miceli
Production/Manufacturing Director: Janet Yearian
Production/Manufacturing Coordinator: Claire Flaherty
Design Manager: John F. Kelly
Photo Editor: Roberta Spieckerman
Design: PCI, San Antonio, TX
Composition: London Road Design, Palo Alto, CA
Illustrations: Pauline Phung, Margaret Copeland, Ray Godfrey
Cover: Ray Godfrey

Photo Acknowledgements: 5 © Ron Sanford/Tony Stone Images; 23 © Owen Franken/Stock, Boston; 27 © Frank Siteman/Stock, Boston; 41 © Joseph Schuyler/Stock, Boston; 45 © Lionel Delevingne/Stock, Boston; 47 © Bert Sagara/Tony Stone Images; 54 (Big Ben) © George Hunter/Tony Stone Images; 54 (Eiffel Tower) © John Lawrence/Tony Stone Images

Order number 21473
ISBN 1-57232-178-4

9 10-BA-01 00

The Connected Mathematics Project Staff

Project Directors

James T. Fey
University of Maryland

William M. Fitzgerald
Michigan State University

Susan N. Friel
University of North Carolina at Chapel Hill

Glenda Lappan
Michigan State University

Elizabeth Difanis Phillips
Michigan State University

Project Manager

Kathy Burgis
Michigan State University

Technical Coordinator

Judith Martus Miller
Michigan State University

Curriculum Development Consultants

David Ben-Chaim
Weizmann Institute

Alex Friedlander
Weizmann Institute

Eleanor Geiger
University of Maryland

Jane Miller
University of Maryland

Jane Mitchell
University of North Carolina at Chapel Hill

Anthony D. Rickard
Alma College

Collaborating Teachers/Writers

Mary K. Bouck
Portland, Michigan

Jacqueline Stewart
Okemos, Michigan

Graduate Assistants

Scott J. Baldridge
Michigan State University

Angie S. Eshelman
Michigan State University

M. Faaiz Gierdien
Michigan State University

Jane M. Keiser
Indiana University

Angela S. Krebs
Michigan State University

James M. Larson
Michigan State University

Ronald Preston
Indiana University

Tat Ming Sze
Michigan State University

Sarah Theule-Lubienski
Michigan State University

Jeffrey J. Wanko
Michigan State University

Evaluation Team

Mark Hoover
Michigan State University

Diane V. Lambdin
Indiana University

Sandra K. Wilcox
Michigan State University

Judith S. Zawojewski
National-Louis University

Teacher/Assessment Team

Kathy Booth
Waverly, Michigan

Anita Clark
Marshall, Michigan

Julie Faulkner
Traverse City, Michigan

Theodore Gardella
Bloomfield Hills, Michigan

Yvonne Grant
Portland, Michigan

Linda R. Lobue
Vista, California

Suzanne McGrath
Chula Vista, California

Nancy McIntyre
Troy, Michigan

Mary Beth Schmitt
Traverse City, Michigan

Linda Walker
Tallahassee, Florida

Software Developer

Richard Burgis
East Lansing, Michigan

Development Center Directors

Nicholas Branca
San Diego State University

Dianne Briars
Pittsburgh Public Schools

Frances R. Curcio
New York University

Perry Lanier
Michigan State University

J. Michael Shaughnessy
Portland State University

Charles Vonder Embse
Central Michigan University

Special thanks to the students and teachers at these pilot schools!

Baker Demonstration School
Evanston, Illinois

Bertha Vos Elementary School
Traverse City, Michigan

Blair Elementary School
Traverse City, Michigan

Bloomfield Hills Middle School
Bloomfield Hills, Michigan

Brownell Elementary School
Flint, Michigan

Catlin Gabel School
Portland, Oregon

Cherry Knoll Elementary School
Traverse City, Michigan

Cobb Middle School
Tallahassee, Florida

Courtade Elementary School
Traverse City, Michigan

Duke School for Children
Durham, North Carolina

DeVeaux Junior High School
Toledo, Ohio

East Junior High School
Traverse City, Michigan

Eastern Elementary School
Traverse City, Michigan

Eastlake Elementary School
Chula Vista, California

Eastwood Elementary School
Sturgis, Michigan

Elizabeth City Middle School
Elizabeth City, North Carolina

Franklinton Elementary School
Franklinton, North Carolina

Frick International Studies Academy
Pittsburgh, Pennsylvania

Gundry Elementary School
Flint, Michigan

Hawkins Elementary School
Toledo, Ohio

Hilltop Middle School
Chula Vista, California

Holmes Middle School
Flint, Michigan

Interlochen Elementary School
Traverse City, Michigan

Los Altos Elementary School
San Diego, California

Louis Armstrong Middle School
East Elmhurst, New York

McTigue Junior High School
Toledo, Ohio

National City Middle School
National City, California

Norris Elementary School
Traverse City, Michigan

Northeast Middle School
Minneapolis, Minnesota

Oak Park Elementary School
Traverse City, Michigan

Old Mission Elementary School
Traverse City, Michigan

Old Orchard Elementary School
Toledo, Ohio

Portland Middle School
Portland, Michigan

Reizenstein Middle School
Pittsburgh, Pennsylvania

Sabin Elementary School
Traverse City, Michigan

Shepherd Middle School
Shepherd, Michigan

Sturgis Middle School
Sturgis, Michigan

Terrell Lane Middle School
Louisburg, North Carolina

Tierra del Sol Middle School
Lakeside, California

Traverse Heights Elementary School
Traverse City, Michigan

University Preparatory Academy
Seattle, Washington

Washington Middle School
Vista, California

Waverly East Intermediate School
Lansing, Michigan

Waverly Middle School
Lansing, Michigan

West Junior High School
Traverse City, Michigan

Willow Hill Elementary School
Traverse City, Michigan

Contents

Thinking with Mathematical Models

Falloi and his mother want to go on the teeter-totter at the park. Falloi weighs 75 pounds and his mother weighs 150 pounds. How can they sit on the teeter-totter so that it balances?

The pep club surveyed 500 students to find out which of several amounts they would be willing to pay for a spirit-week T-shirt. They found that 400 students would pay 2 dollars, 325 would pay 4 dollars, 230 would pay 6 dollars, 160 would pay 8 dollars, and 100 would pay 10 dollars. How can they use this information to predict how many students would pay 5 dollars?

On the day Chantal was born, her Uncle Charlie used $100 to open a savings account. He plans to give Chantal the money in the account on her tenth birthday. The account earns 8% interest at the end of every year. If Charlie does not deposit or withdraw any money, how much money will be in the account on Chantal's tenth birthday?

People in many professions make important predictions every day. Sometimes millions of dollars, and often people's lives, depend on the accuracy of these predictions. A company may develop a wonderful new product, but without accurate predictions about the number and location of potential customers and the price they would be willing to pay, the company could lose a great deal of money. When designing a bridge, a civil engineer must make accurate predictions about how much weight the bridge can hold and about how it will stand up to strong winds and earthquakes. People often use mathematical models to help them make such predictions. People construct mathematical models by gathering data about a situation and then finding a graph or equation that fits the data.

In this unit, you will find mathematical models for many situations, and you will use your models to answer interesting questions like those on the previous page.

Mathematical Highlights

In this unit, you will learn about ways to model relationships between variables.

● Testing paper bridges to find out how thickness affects strength and then fitting a straight line to your experimental data introduces you to the idea of a linear graph model.

● Using what you know about slope and *y*-intercept, you find equations for linear graph models.

● Using graph models and equation models, you make predictions about values that are not in your data set.

● Testing paper bridges to find out how length affects strength and then fitting a curve to your experimental data introduces you to the idea of a nonlinear graph model.

● Finding and graphing (distance, weight) combinations that balance a teeter-totter and (speed, time) combinations needed to complete a trip shows you that sometimes very different situations have similar graph models.

● Calculating interest on a savings account introduces you to another common nonlinear pattern.

● Drawing graphs to model events in a story and writing stories that match the information in graphs give you practice understanding and interpreting graphs of linear and nonlinear relationships.

Using a calculator

In this unit, you might use a graphing calculator to plot data and to fit lines or curves to the data points. As you work on the Connected Mathematics units, you decide whether to use a calculator to help you solve a problem.

Linear Models

Most bridges are built with frames of steel beams. Steel is very strong, but if you put enough weight on any beam, it will bend or break. The amount of weight a beam can support is related to its thickness and design. To design a bridge, engineers need to understand these relationships thoroughly. Engineers often use scale models to test the strength of their bridge designs.

1.1 Testing Paper Bridges

In this problem, you will do an experiment to test some of the principles involved in building bridges.

Equipment: several 11-inch-by-4-inch strips of paper, two books of the same thickness, a small paper cup, and about 50 pennies

Directions:
*   Make a paper "bridge" by folding up 1 inch on each long side of one of the paper strips.

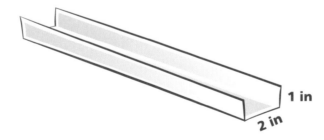

1 in

2 in

- Suspend the bridge between the two books. The bridge should overlap each book by about 1 inch. Place the paper cup in the center of the bridge.

- Put pennies into the cup, one at a time, until the bridge crumples. Record the number of pennies you added to the cup. This number is the *breaking weight* of the bridge.

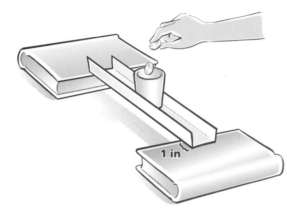

- Put two strips together to make a bridge of double thickness. Find the breaking weight for this bridge. Repeat this experiment to find breaking weights for bridges made from three, four, and five strips of paper.

Problem 1.1

A. Do the experiment described above to find breaking weights for bridges 1, 2, 3, 4, and 5 layers thick.

B. Make a table and a graph of your data.

C. Describe the pattern of change in the data. Then, use the pattern to predict the breaking weights for bridges 6 and 7 layers thick.

D. Suppose you could use half-layers of paper to build the bridges. What breaking weights would you predict for bridges 2.5 layers thick and 3.5 layers thick?

■ Problem 1.1 Follow-Up

How would you expect your results to change if you used a stronger material, such as poster board or balsa wood, to make your bridges?

Drawing Graph Models

A class in Maryland did the bridge-thickness experiment. They combined the results from all the groups and found the average breaking weight for each bridge. They organized their data in a table.

Thickness (layers)	1	2	3	4	5
Breaking weight (pennies)	10	14	23	37	42

The class then made a graph of the data. They thought the pattern looked somewhat linear, so they drew a line to show this trend. This line is a good *model* for the relationship because, for the thicknesses the class tested, the points on the line are close to points from the experiment.

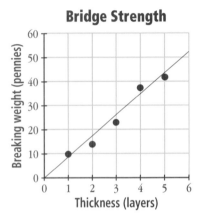

Bridge Strength

The line that the Maryland class drew is a graph model for their data. A **graph model** is a straight line or a curve that shows a trend in a set of data. Once you fit a graph model to a set of data, you can use it to make predictions about values between and beyond the values in your data.

Did you know?

When designing a bridge, engineers need to consider the *load*, or the amount of weight the bridge must support. The *dead load* is the permanent weight of the bridge and fixed objects on the bridge. The *live load* is the weight of moving objects on the bridge. On many metropolitan bridges in Europe—such as the famous Ponte Vecchio in Florence, Italy—dead load can become extremely high as tollbooths, apartments, and even miniature shopping districts are built right onto the surface of the bridge. Live load can usually be controlled by limiting the amount of rail or automobile traffic on the bridge at one time.

A. Draw a straight line that seems to fit the pattern in the (thickness, breaking weight) data you graphed in Problem 1.1.

B. Based on your graph model, what breaking weights would you predict for bridges 6 layers thick and 7 layers thick?

C. Suppose you could use half-layers of paper to build the bridges. What breaking weights would you predict for bridges 2.5 layers thick and 3.5 layers thick?

■ **Problem 1.2 Follow-Up**

1. It is unlikely that your (thickness, breaking weight) data fit a linear pattern exactly. Below are two students' attempts to draw lines to model their group's data.

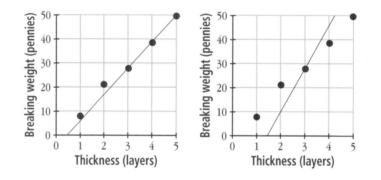

a. Which graph model would allow the students to make better predictions about breaking weights for bridges of different thicknesses? Why?

b. Do you have any suggestions for how the students could change the graph model you chose in part a to help them make even better predictions?

2. For each graph on Labsheet 1.2, try to find a graph model that fits the experimental data as closely as possible. Compare your graph models with those drawn by others in your group. What strategies did you use to help you draw an appropriate graph model?

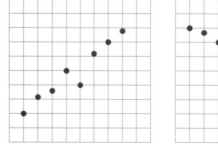
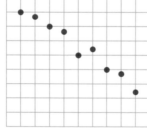
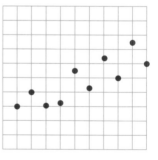

1.3 Finding Equation Models

In your earlier work, you saw that linear relationships can be described by equations of the form $y = mx + b$, where m is the slope, and b is the y-intercept. The line drawn to model the data in the Maryland students' bridge-thickness experiment is shown below. By looking at the graph and comparing vertical change to horizontal change, the students found that the slope of the line is about 8.7. Since the line passes through the origin, its y-intercept is 0. They wrote the equation $y = 8.7x$ to represent the line.

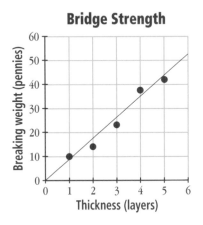

Bridge Strength

Because the equation and the line represent the same information, the equation is also a model of the pattern in the data. This means that you can show the trend in the bridge-thickness data by using a *graph model* or an *equation model*.

Problem 1.3

Use a graphing calculator to explore the equation $y = 8.7x$.

A. Make a table of (x, y) data for $y = 8.7x$, using the x values 0, 1, 2, 3, . . . , 10. Explain how the entries in your table relate to the fact that $m = 8.7$ and $b = 0$.

B. Make a graph of $y = 8.7x$ for $x = 0$ to $x = 10$. Explain how the slope and the y-intercept of the graph are related to the equation.

C. What do the facts that $m = 8.7$ and $b = 0$ mean in terms of bridge thickness and breaking weight?

D. Solve the equation $60 = 8.7x$. What does the solution tell you about bridge thickness and breaking weight?

■ Problem 1.3 Follow-Up

1. Find a linear equation for the graph model from your bridge-thickness experiment. What do the values of m and b in your equation tell you about the graph and about the relationship between bridge thickness and breaking weight for the paper you used?

2. A group in Ms. Hollister's class finds that the equation model $y = 5.5x$ fits their bridge-thickness data. The group conjectures that if they repeat the experiment with cardboard, the relationship will be modeled by $y = 4(5.5x)$.

    a. What is this group's conjecture about the strength of cardboard?

    b. Is $y = 5.5(4x)$ equivalent to $y = 4(5.5x)$? Is $y = 22x$ equivalent to $y = 4(5.5x)$? Give evidence to support your answers.

1.4 Setting the Right Price

To raise money for a local charity, the Student Government Association (SGA) at Kennedy High decides to sell spirit-week baseball caps embroidered with the school mascot. The caps are being donated by a local merchant. The SGA must figure out how many caps to order and how much to charge for each cap.

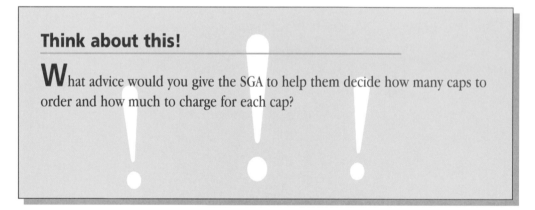

Think about this!

What advice would you give the SGA to help them decide how many caps to order and how much to charge for each cap?

To help with their project, the SGA conducts a market-research study. They ask 500 students in their school which of the following amounts they would be willing to pay for a baseball cap: $2, $4, $6, $8, $10, $12. The results are shown in the table below.

Price (dollars)	2	4	6	8	10	12
Number of buyers	400	325	230	160	100	25

The SGA can use the pattern in the (price, buyers) data to make some predictions.

Problem 1.4

A. Graph the (price, buyers) data, and draw a straight line that models the trend in the data.

B. Write a linear equation of the form $y = mx + b$ for your graph model.

C. What do the patterns of change in the (price, buyers) data and the graph model tell you about the relationship between the price and the number of buyers?

D. What do the values of m and b in the equation model tell you about the relationship between the price and the number of buyers?

E. Which data pair from the survey data is farthest from your graph model? Why do you think this point is so far from the graph?

■ Problem 1.4 Follow-Up

The SGA wrote the equation $y = {}^-40x + 480$ to fit their (price, buyers) data. In this equation, x represents the price, and y represents the number of buyers.

1. What do the numbers $^-40$ and 480 in the equation tell you about the relationship between the price and the number of buyers? What do they tell you about the graph of the relationship?

2. Although the SGA gathered data for only a few prices, they can use their equation model to make a table with information for many more prices.

   a. Make a table like the one below. In the second column, enter the number of buyers the equation model $y = {}^-40x + 480$ predicts for each whole-number price from \$2 to \$15.

Price	Number of buyers	Income
\$2	400	
3	360	
4		
.		
.		
.		

   b. Which data pair from the survey is most different from the prediction made by the equation model? Are any predictions the same as the survey data?

   c. In the third column, enter the income the SGA would earn from each combination of price and buyers.

3. How much should the SGA charge for each cap if they want the maximum possible income? Explain your answer.

4. Make a graph of the (price, income) data. Describe the shape of your graph. How is your answer to question 3 shown on your graph?

1.5 Writing Equations for Lines

Linear equations will not model every relationship you explore. However, you can use them to model relationships in many different situations. For example, Denise and Jonah earn allowances for mowing their lawns each week.

- Denise's father pays her $5 each week.
- Jonah's mother paid him $20 at the beginning of the summer and now pays him $3 each week.

The graphs of the relationships between the number of weeks and the dollars earned are shown below. The graphs were made by plotting points for several values and then fitting lines. Why do you think the lines fit the data points exactly?

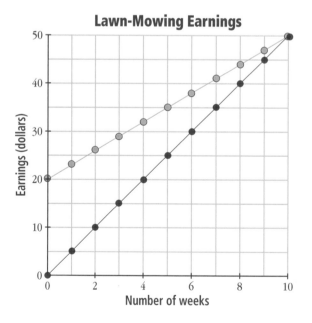

Lawn-Mowing Earnings

Problem 1.5

A. Which graph model shows Jonah's earnings as a function of the number of weeks? In other words, which shows how Jonah's earnings relate to the number of weeks? Which graph model shows Denise's earnings? Explain how you know which graph is which.

B. Write linear equations of the form $y = mx + b$ to show the relationships between Denise's and Jonah's earnings and the number of weeks.

C. What do the values of m and b in each equation tell you about the relationship between the number of weeks and the dollars earned? What do the values of m and b tell you about each graph model?

■ Problem 1.5 Follow-Up

Many problems require you to find a linear equation that matches a pattern in a graph or a table. You know that in a linear equation of the form $y = mx + b$, m is the slope of the line and b is the y-intercept. If you are given the slope and the y-intercept, or if you can read them from a graph or a table, finding an equation is easy. However, finding the slope and the y-intercept often requires a bit of work.

1. How would you find an equation for the line with slope 1.5 that passes through the point (5, 9.5)? Because the slope is the value of m in $y = mx + b$, you know that the equation is of the form $y = 1.5x + b$. But how can you find b?

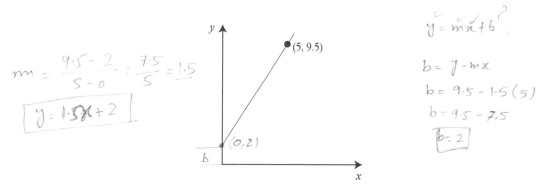

a. Substitute 5 for x and 9.5 for y in the equation $y = 1.5x + b$. Solve for b, and explain what the result tells you about the line. Use your result to help you write an equation for the line.

b. Use your reasoning from part a to find the equation of the line with slope 3 that passes through the point (2, 5).

2. a. Modify the reasoning you used in question 1 to find an equation for the line that passes through the points (2, 1) and (7, 11).

b. Use the reasoning you used in part a to find an equation for the line that passes through the points (1, 8) and (4, 2).

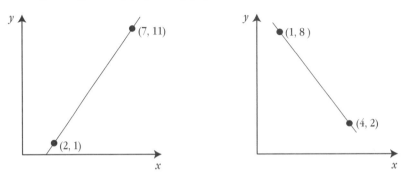

c. Compare the equations you wrote for parts a and b. How are they alike? How are they different?

3. Match each equation with its line in the diagram. Give reasons for your choices.

$x = 1$
$2 = 1$
$31 = 1$
$3x - 1$

a. $y = 0.5x + 3$ $y = 0.5 + 3 = 3.5$ (1, 3.5)
b. $y = {}^-0.5x + 6$ $y = {}^-0.5 + 6 = 5.5$ (1, 5.5)
c. $y = 0.5x + 6$ $y = 0.5 + 6 = 6.5$ (1, 6.5)
d. $y = {}^-0.5x + 3$ $y = {}^-0.5 + 3 = 2.5$ (1, 2.5)

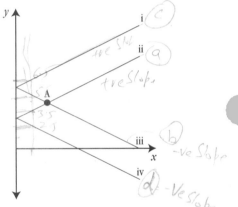

4. a. Point A is on lines ii and iii. Its x-coordinate is 3. Calculate its y-coordinate by substituting 3 for x in the equation for line ii. Then, calculate its y-coordinate by substituting 3 for x in the equation for line iii. What do you observe? Why does this happen?

b. Abdul and Wendy did some calculations to find the y-coordinate of point A. Which student did the calculations correctly? What error did the other student make?

Abdul's work
$y = 0.5 \times 3 + 3$
$y = 0.5 \times 6$
$y = 3$

Wendy's work
$y = 0.5 \times 3 + 3$
$y = 1.5 + 3$
$y = 4.5$

As you work on these ACE questions, use your calculator whenever you need it.

Applications

1. A class in Oregon did the bridge-thickness experiment with construction paper. Their results are shown in this table.

Thickness (layers)	1	2	3	4	5	6
Breaking weight (pennies)	12	20	29	42	52	61

    a. Make a graph of the (thickness, breaking weight) data, and draw a straight line that models the trend in the data.

    b. Use your graph model to estimate the breaking weight for a bridge 7 layers thick.

    c. Find an equation for your graph model, and use it to check your answer for part b.

2. The astronomy club at King Middle School is planning a field trip to the science center to see a new 3-D science film. Renting a bus for the trip will cost $125. Admission to the film is $2.50 per person.

    a. Write an equation to show the relationship between the number of students who go on the trip, *n,* and the cost of the trip, *c.*

    b. Make a table showing the cost of the trip for 0 students, 5 students, 10 students, and so on, up to 50 students.

    c. Make a graph of the equation on grid paper or by using a graphing calculator.

    d. What do the slope and the *y*-intercept of the graph tell you about the cost and the number of students?

3. The student government at Waverly Middle School conducted a survey to find out how much students would be willing to pay to attend a concert by a popular music group. The graph below shows the results of the survey.

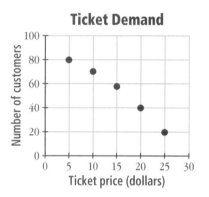

Ticket Demand

a. Make a table showing the approximate (price, customers) data given in the graph.

b. Copy the graph, and draw a straight line that models the trend in the data. Use your model to estimate the number of students who would pay prices of $7.50, $12.50, $17.50, and $22.50.

c. Write an equation for your graph model, and use it to check your estimates from part b.

In 4–8, write an equation for the pattern of change shown in the table.

4.

x	0	1	2	3	4	5	6	7	8
y	0	4	8	12	16	20	24	28	32

5.

x	0	1	2	3	4	5	6	7	8
y	3	8	13	18	23	28	33	38	43

6.

x	1	2	3	4	5	6	7	8	9
y	5	11	17	23	29	35	41	47	53

7.

x	1	3	5	7	9	11	13	15	17
y	2.5	7.5	12.5	17.5	22.5	27.5	32.5	37.5	42.5

8.

x	0	1	2	3	4	5	6	7	8
y	3	2.5	2	1.5	1	0.5	0	⁻0.5	⁻1

In 9–13, find the equation for the line.

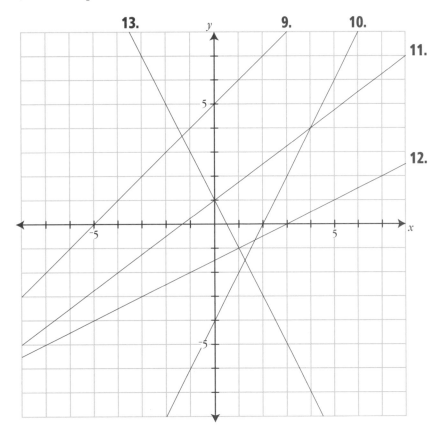

14. Write an equation for the line with slope 4.2 and y-intercept $(0, 3.4)$.

15. Write an equation for the line with slope $\frac{2}{3}$ and y-intercept $(0, 5)$.

16. Write an equation for the line with slope 2 that passes through the point $(4, 12)$.

17. Write an equation for the line that passes through the points $(1, 3)$ and $(7, 6)$.

18. Write an equation for the line that passes through the points (0, 15) and (5, 3).

19. Write an equation for the line that passes through the points (2, 7) and (5, 4).

In 20–23, the coordinates of two points on a line are given. Find an equation for the line, and give the slope and *y*-intercept of the line.

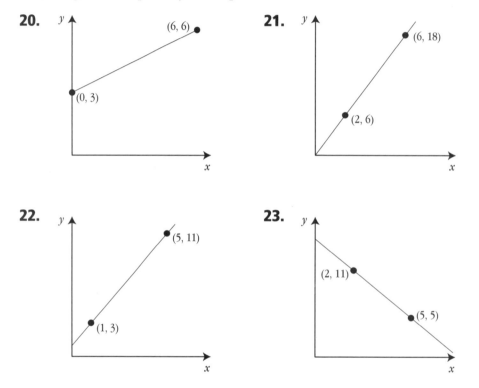

20. (6, 6) (0, 3)

21. (6, 18) (2, 6)

22. (5, 11) (1, 3)

23. (2, 11) (5, 5)

24. Spartan Publishing Company's delivery truck has a broken fuel gauge. Luckily, the driver always keeps track of mileage and gas consumption. She uses her data to write the equation $G = 25 - \frac{1}{15}M$ for the relationship between the number of gallons of gasoline in the tank, *G,* and the number of miles driven since the last fill-up, *M.*

    a. The driver has just filled the tank and is about to start the engine. What is the value of *M* for this situation? Use the equation to figure out the value of *G* for this situation. What does the result tell you about the gas tank?

b. If the driver travels 50 miles after filling the tank, how much gas will be left?

c. After filling the tank, how many miles can the driver travel before 5 gallons remain?

d. Solve the equation $10 = 25 - \frac{1}{15}M$. What does the solution tell you about the amount of gas left and the number of miles driven since the last fill-up?

e. Use the equation to figure out how many miles the driver would have to travel to use 1 gallon of gas. Explain how you got your answer.

f. In the equation $G = 25 - \frac{1}{15}M$, what do the numbers 25 and $\frac{1}{15}$ tell you about the situation?

g. What patterns would you expect to see in the table and the graph of the relationship between miles driven and gallons of gas remaining? How would the numbers 25 and $\frac{1}{15}$ from the equation be represented in the table and the graph?

Connections

25. A survey of one homeroom class at Pioneer Middle School finds that 20 out of 30 students in the class would be willing to spend $8 for a school-spirit T-shirt.

a. If there are 600 students in the school, how many do you predict would be willing to spend $8 for a school-spirit T-shirt?

b. If there are 450 students in the school, how many do you predict would be willing to spend $8 for a school-spirit T-shirt?

26. The table below gives name lengths, heights, and foot lengths for a group of middle school students.

Student	Name length (letters)	Height (cm)	Foot length (cm)
Thomas Petes	11	126	23
Michelle Hughes	14	117	21
Shoshana White	13	112	17
Deborah Locke	12	127	21
Tonya Stewart	12	172	32
Richard Mudd	11	135	22
Tony Tung	8	130	20
Janice Vick	10	134	21
Bobby King	9	156	29
Kathleen Boylan	14	164	28

a. Make a graph of the (name length, height) data, a graph of the (name length, foot length) data, and a graph of the (height, foot length) data.

b. Of the graphs you made in part a, which could be modeled with a straight line? Explain your answer.

c. For each graph you think can be modeled with a straight line, explain what the line would tell you about the relationship between the variables.

27. You can produce odd and even numbers by substituting integers for n in the linear equations $O = 2n + 1$ and $E = 2n$.

a. Copy and complete the table below to show the odd and even numbers produced by these equations for the given values of n.

n	0	1	2	5	10
O					
E					

b. Compare the equations for O and E to the general linear form $y = mx + b$. What are the values of m and b in each equation? What do these values tell you about the patterns you would see in the table and the graphs of the equations?

28. The equations below express the relationship between a person's height, *H*, and femur (thigh bone) length, *F*. Both measurements are in inches.

$$\text{Males: } H = 27.5 + 2.24F$$
$$\text{Females: } H = 24 + 2.32F$$

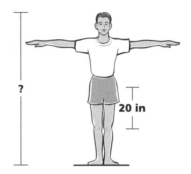

a. If a man's femur is 20 inches long, about how tall is he?

b. If a woman's femur is 15 inches long, about how tall is she?

c. If a man is 65 inches tall, about how long is his femur?

d. If a woman is 68 inches tall, about how long is her femur?

e. What do the numbers 2.24 and 2.32 tell you about the relationship between femur length and height for males and females?

f. What do the numbers 2.24 and 2.32 tell you about the patterns you would see in the tables and the graphs of (femur length, height) data?

In 29–31, tell whether the data in the table represent a linear relationship, and explain your answer.

29.

x	2	4	6	8	10	12	14
y	0	1	2	3	4	5	6

30.

x	1	2	3	4	5	6	7
y	0	3	8	15	24	35	48

31.

x	1	4	6	7	10	12	16
y	2	⁻1	⁻3	⁻4	⁻7	⁻9	⁻13

In 32–38, tell whether the equation represents a linear relationship, and explain how you arrived at your conclusion. If the equation is linear, give the slope and the y-intercept of the graph.

32. Renting a bus for a school trip costs $240, and food and accomodations cost $25 per person. If n people go on the trip, the cost per person, C, in dollars is

$$C = \frac{240 + 25n}{n}$$

33. At Josie's Buffet, lunch costs $6.50 per person. The total cost, T, in dollars for n people to eat lunch is $T = 6.5n$.

34. If the cost for car insurance is d dollars per year, then the cost, C, in dollars for a three-month policy is $C = \frac{d}{4}$.

35. $y = \frac{20}{x}$ **36.** $y = 20x$ **37.** $y = \frac{x}{20}$ **38.** $y = 0.05x$

Extensions

39. The 24 students in Mr. McKeever's homeroom were asked which of several prices they would be willing to pay for a ticket to the school play. The results are shown in this table.

Ticket price (dollars)	1.00	1.50	2.00	2.50	3.00	3.50	4.00	4.50
Probable ticket sales	20	20	18	15	12	10	8	7

a. There are 480 students in the school. Use the data from Mr. McKeever's class to predict ticket sales for the entire school for each price in the table.

b. Use your results from part a to predict the income the school could expect if tickets were sold at each price in the table.

c. What ticket price should the school charge if they want the maximum income?

40. This table gives the numbers of stories and the heights for 15 U.S. buildings.

Building	Location	Number of stories	Height (feet)
National City Center	Akron, Ohio	23	301
IBM Tower	Atlanta, Georgia	50	813
U.S.F. & G. Building	Baltimore, Maryland	40	529
John Hancock Tower	Boston, Massachusetts	60	800
NCNB Plaza	Charlotte, North Carolina	40	503
Sears Tower	Chicago, Illinois	110	1454
Sohio Tower	Cleveland, Ohio	46	650
Interfirst Plaza	Dallas, Texas	73	939
Westin Hotel	Detroit, Michigan	71	720
Pacific Tower	Honolulu, Hawaii	30	350
Texas Commerce Tower	Houston, Texas	75	1002
Indiana Bell Telephone Bldg.	Indianapolis, Indiana	20	320
Los Angeles City Hall	Los Angeles, California	28	454
Empire State Building	New York, New York	102	1250
United Nations Building	New York, New York	39	505

a. Graph the (stories, height) data for these buildings, and draw a straight line that models the trend in the data.

b. Write an equation of the form $y = mx + b$ for your graph model, and explain what the value of m tells you about this situation.

c. The Mellon Bank Center in Philadelphia, Pennsylvania, has 56 stories and is 880 feet tall. How does this data pair compare to the data pair predicted by your equation model from part b?

41. This table gives the average heights of boys of even-numbered ages from birth to age 16.

Age (years)	0	2	4	6	8	10	12	14	16
Height (cm)	52	87	103	116	127	138	150	163	174

a. Graph the (age, height) data, and draw a straight line that models the trend in the data.

b. Write an equation of the form $y = mx + b$ for your graph model. What do the values of m and b tell you about age and height in this situation?

c. Use your model to estimate the average heights of boys of odd-numbered ages from 1 to 15. Then, compare your estimates to the actual data given in the table below.

Age (years)	1	3	5	7	9	11	13	15
Height (cm)	74	91	107	119	132	142	152	163

Mathematical Reflections

In this investigation, you used graph models and equation models to represent linear relationships between variables. These questions will help you summarize what you have learned:

1 In Problem 1.2, you drew a straight line to model the trend in the (thickness, breaking weight) data. For data that can be modeled with a straight line, how do the y values change as the x values increase or decrease?

2 Write the general form of the equation for a linear relationship. Explain what each part of the equation means.

3 If you know the coordinates of two points on a line, how can you find an equation for the line? Use an example if it helps you to explain your thinking.

4 **a.** Describe the graph of a line that has a negative slope.

   b. What part of the linear equation $y = mx + b$ tells you whether the graph has a negative slope?

   c. For a line with a negative slope, how do the y values change as the x values increase?

5 **a.** Explain how you would check whether the point (2, 7.5) is on the line with equation $y = 3x + 0.5$.

   b. After you substitute a number for x in the equation $y = 10x + 2.8$, in what order should you do the calculations to find the value of y? What does the result of your calculations tell you?

Think about your answers to these questions, discuss your ideas with other students and your teacher, and then write a summary of your findings in your journal.

Nonlinear Models

In the last investigation, you tested paper bridges of various thicknesses. You found that thicker bridges are stronger than thinner bridges, and you discovered that the relationship between thickness and breaking weight is approximately linear. In this investigation, you will explore relationships with different types of models.

2.1 Testing Bridge Lengths

In this problem, you will experiment with paper bridges of various lengths. What relationship do you expect to find between the length of a bridge and its breaking weight? Do you think longer bridges will be stronger or weaker than shorter bridges?

Equipment: eight 4-inch-wide strips of paper with lengths 4, 5, 6, 7, 8, 9, 10, and 11 inches, two books of the same thickness, a small paper cup, and about 50 pennies

Directions:

• Make paper bridges by folding up 1 inch on each long side of the paper strips.

• Suspend one of the bridges between the two books. The bridge should overlap each book by about 1 inch. Place the paper cup in the center of the bridge.

• Put pennies into the cup, one at a time, until the bridge crumbles. Record the length and breaking weight of the bridge.

• Repeat this experiment to find the breaking weights of the remaining bridges.

Problem 2.1

A. Do the experiment described to find the breaking weights of paper bridges of lengths 4, 5, 6, 7, 8, 9, 10, and 11 inches. Organize your data in a table. Study the table, and look for a pattern. Do you think the relationship between bridge length and breaking weight is linear?

B. Make a graph of the (length, breaking weight) data from your experiment, and describe the pattern you see. Do the data appear to be linear?

C. Draw a straight line or a curve that seems to model the trend in the data. Do you think your graph fits the data satisfactorily? Explain.

D. Use your graph model to predict breaking weights for bridges of lengths 4.5, 5.5, and 6.5 inches. Make bridges of these lengths, and test your predictions.

E. How is the relationship between bridge length and breaking weight in this problem similar to and different from the linear relationships you studied in the last investigation?

■ Problem 2.1 Follow-Up

1. If you were to do the bridge-length experiment using strips of paper 12, 13, 14, and 15 inches long, what pattern would you expect to see in the results? Explain your reasoning.

2. What breaking weight would you expect for a bridge 3 inches long? Explain your reasoning.

Did you know?

When designing bridges, engineers must think about *impact*, or the effect of outside forces on the bridge. The design must account for gusting winds, shifting water, and the possibility of seismic activity. In addition, engineers must consider changes in air temperature that may cause the materials in the bridge to expand or contract. Typically, the milder temperature ranges of coastal regions create less of a change in size than the more extreme temperature ranges of desert and mountain areas.

2.2 Keeping Things Balanced

The pattern you saw in the (length, breaking weight) data shows up in many other important and familiar problems. For example, you may have thought about this problem when you were a child:

How do people of different weights balance on a teeter-totter?

You probably discovered that the lighter person has to sit farther from the balance point (called the *fulcrum*) than the heavier person does.

| fulcrum | fulcrum |

You might have found the balance point by trial and error, but there is a mathematical relationship between weight and distance. The following experiment will help you discover this relationship.

Equipment: a meterstick, several identical small weights (heavy washers work well), and a block of wood to use as a fulcrum (a triangular prism works best)

Directions:
• Balance the meterstick on the fulcrum. The 50-centimeter mark should be directly over the fulcrum.

• Place 3 weights 40 centimeters to the left of the fulcrum. Leave these weights in this position for the entire experiment.

• Place 4 weights to the right of the fulcrum so that the meterstick balances. Record the distance of the 4 weights from the fulcrum.

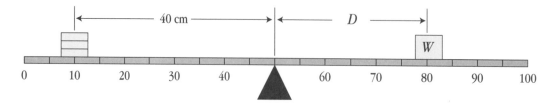

- Find several other (distance, weight) combinations that balance the meterstick. For example, where would you have to place 6 weights? 10 weights? How many weights would you have to place 10 centimeters from the fulcrum? How many weights would you have to place 20 centimeters from the fulcrum?

Problem 2.2

A. Do the experiment described above, and make a table of the (distance, weight) combinations you find. What does your table suggest about the relationship between distance and weight?

B. Make a graph of your data, and draw a straight line or a curve that models the trend. What does your graph suggest about the relationship between distance and weight?

C. How is the pattern of change in the (distance, weight) data similar to and different from the pattern of change in the (bridge length, breaking weight) data from Problem 2.1?

D. How is the pattern of change in the (distance, weight) data similar to and different from the pattern of change in the (bridge thickness, breaking weight) data from Problems 1.1 and 1.2?

Problem 2.2 Follow-Up

1. How would the results of this experiment have been different if you had started with 2 weights placed 30 centimeters to the left of the fulcrum? With 4 weights placed 30 centimeters to the left of the fulcrum? With 2 weights placed 50 centimeters to the left of the fulcrum?

2. Some (distance, weight) combinations that balance 3 weights 40 centimeters from the fulcrum are (30, 4), (24, 5), and (20, 6). Which of the following equations do these three data pairs satisfy? Explain.

a. $W = 120D$

b. $W = \frac{120}{D}$

c. $D = \frac{120}{W}$

d. $W = \frac{D}{120}$

e. $WD = 120$

f. $DW = 120$

Testing Whether Driving Fast Pays

Riders on the Ocean and History Bicycle Tour cycle from Philadelphia, Pennsylvania, along the Atlantic Coast to Norfolk, Virginia, and then to Williamsburg, Virginia. When the tour is over, the riders take a bus from Williamsburg back to Philadelphia—a distance of about 300 miles. Everyone is anxious to get home, so the bus driver is tempted to drive fast.

In this problem, you will study the relationship between the average speed of the bus and the time it takes to reach Philadelphia. You will see how much difference in travel time driving fast really makes.

Problem 2.3

A. Copy and complete the table below to show the time it would take for the 300-mile trip at various average speeds.

Average speed (miles per hour)	30	40	50	60	70
Trip time (hours)					

B. Make a graph of the relationship between the average speed, S, and the time, T.

C. Find an equation for the relationship between S and T.

D. Is the relationship between S and T linear or nonlinear? Explain how the table, the graph, and the equation support your answer.

■ **Problem 2.3 Follow-Up**

1. The bus driver figured out that if he increased his average speed from 40 to 45 miles per hour, the time for the trip would be shortened from $7\frac{1}{2}$ hours to $6\frac{2}{3}$ hours, a savings of 50 minutes. He then reasoned that increasing his average speed from 45 to 50 miles per hour would cut another 50 minutes off the trip, and increasing his average speed from 50 to 55 miles per hour would cut another 50 minutes off the trip.

  a. Do you agree with the bus driver's conclusions about the time he would save by driving faster? Explain your reasoning.

  b. How is your answer for part a illustrated in your graph of the (speed, time) data?

2. How would the table, the graph, and the equation change if the trip were 500 miles instead of 300 miles?

3. Look back at your work from Problem 2.2. Find an equation for the relationship you explored in that problem.

As you work on these ACE questions, use your calculator whenever you need it.

Applications

1. These data are from an ad for a construction crane. They show the maximum weight the crane arm can lift at various distances from the cab out along the arm.

Distance (ft)	12	24	36	48	60
Weight (lbs)	7500	3750	2500	1875	1500

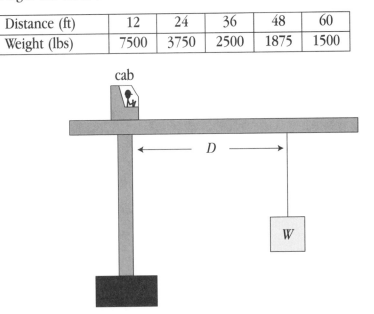

cab

D

W

a. Describe the relationship between distance and weight for the crane.

b. Write an equation for the weight, W, as a function of the distance, D.

c. Use your equation to estimate the weight the crane can lift at these distances from the cab.

   i. 18 ft   **ii.** 30 ft   **iii.** 72 ft

d. Make a graph of the relationship between D and W, and explain how the graph's shape illustrates the relationship you described in part a.

e. How, if at all, is this situation related to the bridge-length and teeter-totter experiments?

2. Suppose you and some friends are planning a 20-mile bike trip to a park campsite.

a. At what average speed would you have to ride to complete the trip in 4 hours? In 2 hours? In 1.5 hours? In 1 hour?

b. Write an equation for the relationship between riding time, t, in hours and average speed, s, in miles per hour for a 20-mile trip.

c. Is the relationship between t and s linear? Use a table and a graph to support your answer.

d. Which of the following causes the greater change in required average speed: increasing the riding time from 1 hour to 2 hours or increasing the riding time from 2 hours to 3 hours?

In 3–6, find the value of W or D that will balance the teeter-totter.

3.

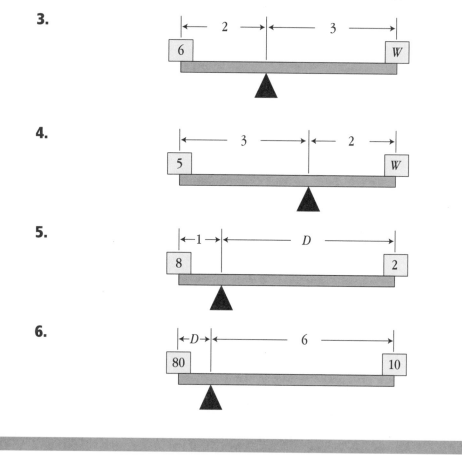

4.

5.

6.

Connections

7. Find all combinations of whole-number values for D and W that will make this teeter-totter balance.

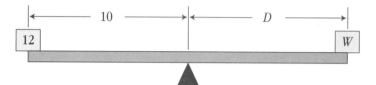

8. **a.** At right is a scale drawing of a rectangle with an area of 300 ft^2. Make scale drawings of some other rectangles with areas of 300 ft^2.

    b. What width must a rectangle with an area of 300 ft^2 have if its length is 1 ft? 2 ft? 3 ft? L ft?

    c. How does the width change as the length increases?

    d. Make a graph of (W, L) pairs that give an area of 300 ft^2. Explain how your graph illustrates your answer for part c.

    e. Is the relationship between W and L linear or nonlinear? Explain how the pattern in your graph can help you answer this question.

9. **a.** The rectangle pictured in question 8 has a perimeter of 70 ft. Make scale drawings of some other rectangles with perimeters of 70 ft.

    b. What width must a rectangle with a perimeter of 70 ft have if its length is 1 ft? 2 ft? L ft?

    c. How does the width change as the length increases?

    d. Make a graph of (W, L) pairs that give a perimeter of 70 ft. Explain how your graph illustrates your answer for part c.

    e. Is the relationship between W and L linear or nonlinear? Explain how the pattern in your graph can help you answer this question.

10. On the Ocean and History Bicycle Tour, there are some free mornings when participants can choose from several side trips.

   a. The time allowed for a side trip is 4 hours. What (distance, average speed) combinations are possible for a side trip? Record some of the possible combinations in a table.

Distance (miles)	20				
Average speed (miles per hour)	5				

   b. Write an equation showing the relationship between the average speed and the distance.

   c. Use your table or equation to find the average speed required for a 60-mile trip.

   d. Use your table or equation to find the distance a helicopter traveling at an average speed of 80 miles per hour could cover in 4 hours.

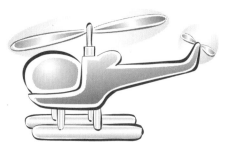

   e. Is the relationship between distance and average speed linear? Explain how you found your answer.

11. In this problem, you will use a graphing calculator to explore graphs of $y = a + \frac{2}{x}$ for different positive values of a.

   a. Set the window on your calculator to display x and y values from $^-8$ to 8. Graph $y = \frac{2}{x}$. Make a sketch of the graph.

   b. Graph each of these equations, and watch what happens to the graphs as different positive numbers are added to the value of y:

$$y = 1 + \frac{2}{x} \qquad y = 2 + \frac{2}{x} \qquad y = 3 + \frac{2}{x} \qquad y = 4 + \frac{2}{x}$$

   c. Describe what happens to the graph of $y = \frac{2}{x}$ when a positive number is added to the value of y.

   d. Describe what the graph of $y = 100 + \frac{2}{x}$ would look like.

Extensions

12. In this problem, you will use a graphing calculator to explore graphs of $y = a + \frac{2}{x}$ for different negative values of a.

    a. Set the window on your calculator to display x and y values from $^-8$ to 8. Graph $y = \frac{2}{x}$. Make a sketch of the graph.

    b. Graph each of these equations, and watch what happens to the graphs as different negative numbers are added to the value of y:

$$y = {}^-1 + \frac{2}{x} \qquad y = {}^-2 + \frac{2}{x} \qquad y = {}^-3 + \frac{2}{x} \qquad y = {}^-4 + \frac{2}{x}$$

    c. Describe what happens to the graph of $y = \frac{2}{x}$ when a negative number is added to the value of y.

    d. Describe what the graph of $y = {}^-100 + \frac{2}{x}$ would look like.

Mathematical Reflections

In this investigation, you explored nonlinear relationships.

- In Problem 2.1, you investigated the relationship between length and breaking weight for a bridge.

- In Problem 2.2, you investigated the relationship between weight and distance from the fulcrum for a teeter-totter.

- In Problem 2.3, you investigated the relationship between speed and time for a trip.

These questions will help you summarize what you have learned:

1 What common patterns of change did you find in the tables and the graphs of these relationships? Did all three relationships fit the same pattern?

2 How are the patterns in the tables and the graphs of these relationships different from the patterns in the tables and the graphs of the linear relationships you explored in Investigation 1?

3 There are many types of nonlinear relationships, some of which you will explore later this year. The type of relationship you studied in this investigation is called an *inverse* relationship. Can you think of some reasons this name is appropriate?

Think about your answers to these questions, discuss your ideas with other students and your teacher, and then write a summary of your findings in your journal.

More Nonlinear Models

In previous investigations, you found graph models and equation models to fit trends in data. You saw that finding models for data can help you make predictions. In Investigation 1, you worked with data that showed linear trends. You drew straight lines to model the data and then wrote equations for the lines. The data in Investigation 2 showed nonlinear trends. You may have found it difficult to find graph and equation models for these data.

In this investigation, you will continue to explore nonlinear situations. You will begin to see familiar patterns in tables and graphs of data. For example, you saw that the data pattern for the teeter-totter experiment was similar to the data pattern for the bridge-length experiment. Recognizing a familiar pattern can give you hints about what type of model might fit a set of data.

As you work through the remaining problems in this book, think about how the data patterns you find compare with patterns you have seen before.

3.1 Earning Interest

On the day Chantal was born, her Uncle Charlie used $100 to open a savings account. He planned to use the money in the account to buy Chantal a gift on her tenth birthday. Charlie forgot about the money until last week when he received an invitation to Chantal's fourteenth birthday party.

Charlie called Chantal and told her about the account. He said, "I'm sorry I forgot about your tenth birthday. Would you like me to give you the money that's in the account now, or would you like me to leave it in the bank and give it to you on your eighteenth birthday?"

The savings account that Charlie opened earns 8% interest each year. This means that at the end of each year, the bank calculates 8% of the balance and adds this amount to the account. At the end of the first year, the interest earned was

$$I = 8\% \text{ of } \$100$$
$$I = 0.08 \times \$100$$
$$I = \$8$$

The interest was added to the account, giving a new balance of $108. So, on Chantal's first birthday there was $108 in the account.

At the end of the second year, the bank again calculated 8% interest and added it to the account. Since there was $108 in the account, the interest earned was

$$I = 8\% \text{ of } \$108$$
$$I = 0.08 \times \$108$$
$$I = \$8.64$$

This amount was added to the account, giving a new balance of $108 + $8.64 = $116.64. So, on Chantal's second birthday there was $116.64 in the account.

Problem 3.1

Charlie's account has been earning 8% interest at the end of every year since Chantal was born. Charlie has not deposited or withdrawn any money since he opened the account. How much money is in the account on Chantal's fourteenth birthday?

1. Copy and complete the table to show the birthday year, the balance at the beginning of that year, and the interest earned at the end of that year. Your table should include data up through Chantal's fourteenth birthday.

Birthday	Balance	Interest earned
0 (birth)	$100.00	$8.00
1	108.00	8.64
2	116.64	
3		
.		
.		
.		

2. Make a graph of the (birthday, balance) data from your table, and draw a straight line or curve to model the trend. Is the graph linear? Explain how you know.

3. About how much money will be in the account on Chantal's eighteenth birthday? Explain how you found your answer.

4. Do you think Chantal should take the money in the account now or wait until her eighteenth birthday? Explain your answer.

5. Is the pattern of change for these data similar to the pattern in the bridge-thickness data, the bridge-length data, or the teeter-totter data? Explain.

3.2 Pouring Water

When you describe a relationship between two variables, you might say that as the x values increase, the y values also increase or that as the x values increase, the y values decrease. However, such a description does not tell the whole story.

In both the graph of the (thickness, breaking weight) data and the graph of the (birthday, balance) data, an increase in the x values produces an increase in the y values. However, the shapes of these graphs are very different. In the (thickness, breaking weight) graph, the y values increase at a constant rate. In the (birthday, balance) graph, the rate at which the y values increase is not constant.

In this problem, you will experiment with another relationship. Think about how the graph of this relationship compares with other graphs you have seen.

As a class, conduct the following experiment:

Equipment: 8 identical clear drinking glasses, stick-on notes or masking tape, a pen or marker, and water

Directions:
- Line up the glasses. Use stick-on notes or masking tape to label the glasses 1, 2, 3, 4, and so on.

- Fill glass 1 with water.

- Pour half the water from glass 1 into glass 2, and return glass 1 to its place.

- Pour half the water from glass 2 into glass 3, and return glass 2 to its place.

- Continue this process until you have poured half the water from glass 7 into glass 8.

Look at the water levels in the glasses. How is the glass number related to the amount of water in the glass?

Problem 3.2

A. If you started the experiment with 32 ounces of water in glass 1, how much water would be in each glass when you were done pouring? Arrange these (glass number, amount of water) data into a table. Describe the relationship between the glass number and the amount of water in the glass.

B. Make a graph of the data, and draw a straight line or a curve to model the trend.

C. Describe the pattern you see in your graph.

D. If you continued to add glasses and pour half the water from the last glass into the new glass, how much water would be in glass 20?

■ Problem 3.2 Follow-Up

1. Does it make sense to connect the data points on your graph? Why or why not?

2. Compare the graph you made in Problem 3.2 with the other graphs you have made in this unit. Describe how it is similar to and different from the other graphs.

3. If you continued to add glasses and pour half the water from the last glass into the new glass, would you ever run out of water? Explain your answer.

As you work on these ACE questions, use your calculator whenever you need it.

Applications

1. Betty's Bakery sells giant cookies for $1.00 each. This price is no longer high enough to create a profit, so Betty decides to raise the price. She doesn't want to shock her customers by raising the price too suddenly or too dramatically. She considers these three plans:

- Plan 1: Raise the price by $0.05 each week until the price reaches $1.80.

- Plan 2: Raise the price by 5% each week until the price reaches $1.80.

- Plan 3: Raise the price by the same amount each week for 8 weeks, so that in the eighth week the price reaches $1.80.

   a. Make a table for each plan. How many weeks will it take the price to reach $1.80 under each plan?

   b. On the same set of axes, graph the data for each plan. Compare the shapes of the graphs and what they mean in terms of the changing cookie price paid by customers.

   c. Are any of the graphs you drew linear? Explain.

   d. Which plan do you think Betty should implement? Give reasons for your choice.

2. Betty, the owner of Betty's Bakery, suspects that someone is stealing her chocolate chips.

    a. There are 1 million chips in a new canister of chocolate chips. Betty uses about 40,000 chips each day. How many days should the canister last?

    b.   **i.** Make a graph that shows the relationship between the number of days after Betty opens a new canister and the approximate number of chips that should be in the canister at the end of each day.

        ii. Write an equation for this relationship.

    c. A gauge on the side of the chip canister allows Betty to estimate the number of chips remaining. On a day when she opens a new canister, Betty begins keeping track of the approximate number of chips left at the end of each day.

Day	1	2	3	4	5	6	7	8
Chips left	800,000	640,000	512,000	410,000	330,000	260,000	210,000	170,000

        Make a graph of these data. Compare this graph with your graph from part b. Are Betty's suspicions about the chocolate chip thefts justified? Explain.

Connections

3. Since Betty raised her prices, cookie sales have fallen. Betty calls in a business consultant to help. The consultant suggests that Betty conduct a customer survey. Betty's customers are asked which of several amounts they would be willing to pay for a cookie. Here are the results:

Price	$1.75	$1.50	$1.25	$1.00
Customers willing to pay this price	100	117	140	175

    a. Make a graph of these data, and draw a straight line or a curve that models the trend.

    b. Use your graph model to predict the number of customers who would be willing to pay $1.35 and the number who would be willing to pay $2.00.

    c. Do you think predictions based on your graph model are accurate? Explain.

d. The shape of this graph resembles the shape of another graph you have drawn. Look back at your work in this unit. Which situation has a graph similar to this one?

4. When you see a trend in plotted data, you can usually draw a graph model that you can use to make predictions. Labsheet 3.ACE contains the data graphs shown below. For each graph, decide whether there is a trend in the data. If there is a trend, do the following:

- Draw a straight line or a curve that models the trend. If the graph model is a line, give its slope.

- Identify a situation you studied in this unit that has a similar graph model.

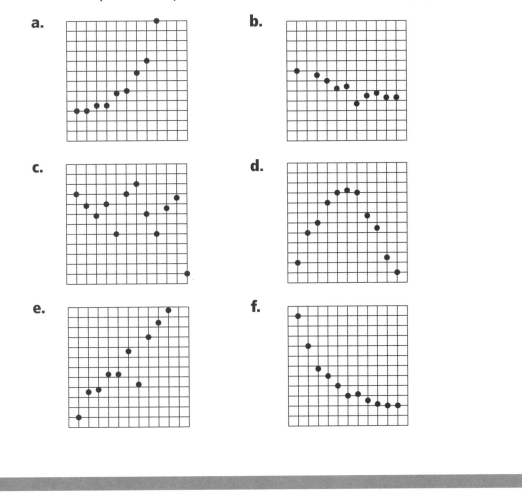

5. Betty's Bakery has a phone system for which the monthly bill depends on the amount of time the phones are in use. The billing formula is $c = 50 + 6h$, where c is the cost in dollars and h is the number of hours of use.

    a. In June, the phone was used for 70 hours. What was the phone bill for June?

    b. What do the numbers 50 and 6 in the billing formula tell you about the relationship between the hours of use and the monthly bill?

    c. Is the relationship between cost and hours of use linear or nonlinear?

    d. What do the numbers 50 and 6 tell you about the pattern you would see in a table and a graph of the (hours, cost) data?

    e. Solve the equation $230 = 50 + 6h$, and explain what the answer tells you about the phone bill for Betty's Bakery.

Extensions

6. Kareem's youth group is going on an outing. They will rent a dormitory for $600 and charter a bus for $250. In addition, each person will pay $20 for food and supplies.

    a. In general, how will the cost per person change as the number of people increases?

    b. Copy and complete this table to show the cost per person for various numbers of people.

Number of people	5	10	15	20	25	30
Cost per person (dollars)						

    c. Make a graph of the data, and draw a straight line or a curve that models the trend. Describe the pattern of change.

    d. Use your graph to predict the cost per person if 37 people go on the trip.

    e. Write an equation for the relationship between the number of people, n, and the cost per person, c.

f. Use your equation to predict the cost per person if 37 people go on the trip.

g. Compare the predictions given by the graph and equation models. Which prediction do you think is most accurate? Explain your choice.

h. Which problems from this unit have graphs with similar patterns of change?

7. Four biologists are studying the caribou and wolf populations in a particular area of Alaska. The caribou are prey to the wolves, and keeping the two populations in balance is important to the survival of both species. The biologists are trying to predict what will happen if no measures are taken to control the populations. After studying

the situation, each biologist makes a graph of his or her prediction. Describe what each graph represents in terms of the animal populations.

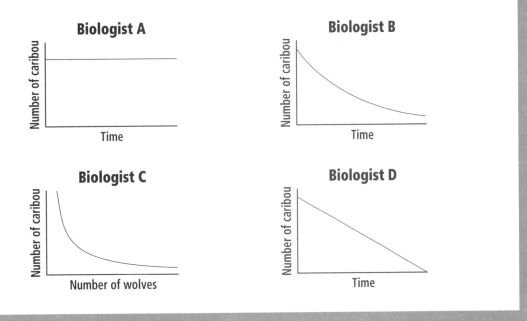

Mathematical Reflections

In this investigation, you explored more nonlinear relationships.

- In Problem 3.1, you investigated the relationship between the number of years money is invested and the amount of money in the bank account.

- In Problem 3.2, you investigated the relationship between the glass number and the amount of water in the glass.

These questions will help you summarize what you have learned:

1 How are the graph models for these two relationships similar?

2 How are the graph models for these two relationships different?

3 How can you tell that a relationship is linear without making a graph?

4 Describe the different kinds of graph models you have discovered in your work so far in this unit.

Think about your answers to these questions, discuss your ideas with other students and your teacher, and then write a summary of your findings in your journal.

A World of Patterns

In this Investigation, you will sketch graphs that fit written descriptions, and you will make up stories about what a given graph might represent.

4.1 Modeling Real-Life Events

The East Coast bus tour is just entering Charleston, South Carolina. The tour group plans to see a warship from the Civil War era and then have lunch. When they arrive at the ship, the tide is low and the ship is nearly aground in the harbor. They are told that no more tours may board the ship until the tide rises and the ship is afloat. The tour cannot agree on what to do while they wait, so they divide into smaller groups.

One group goes for an early lunch. Unfortunately, the furnace at the restaurant breaks just as they arrive. It's a cold day, and the restaurant cools down very quickly until it reaches the outside temperature. The manager apologizes and asks the group to come back later for a free meal.

A second group goes parasailing. It's cold outside, so they wear wet suits. Paige goes first. She is pulled along briefly by the boat. Then her parachute opens, and she soars into the air. After taking her once around the harbor, the boat operator adjusts the speed, and Paige descends gradually and lands gently on a platform at the back of the boat. The others take their turns, and they all agree that the experience is thrilling.

A third group decides to get some exercise after sitting so long on the bus. They rent bicycles and ride at a steady pace along some quiet, winding streets. By the time they finish their ride, the tide has risen, and they meet the rest of the group for the boat tour. After the boat tour, everyone goes out to dinner and then walks along the beach as the tide goes out.

Many of the events in the story of the group's day can be modeled with graphs. In this problem, you will explore some of these graph models.

Problem 4.1

A. The graphs below model events described in the story. Tell which event each graph represents, and explain why you think the graph is a good model for the event. Copy each graph, and label the axes with the variable names.

1.　　　　　　　　　　　**2.**

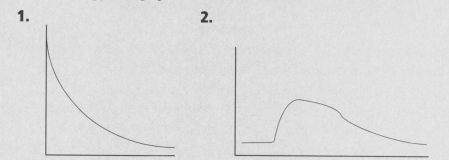

B. Identify other events in the story that involve relationships between two variables. For each event, sketch a graph that models the relationship between the variables. Carefully label the axes so it is easy to see what the graph represents.

C. For each graph you sketched in part B, write a sentence or two explaining what the graph shows and how it fits the description in the story.

D. Write a paragraph about the experiences a fourth group had while waiting for the tide to rise. Sketch a graph or graphs that show how the variables in your story are related.

■ Problem 4.1 Follow-Up

The following graphs show the relationship between time and distance traveled by a cyclist.

1. a. Describe the relationship shown in this graph. Make up a story about the cyclist that could be modeled by the graph.
  b. Over what interval in the graph is the cyclist traveling the fastest? Explain how you know.

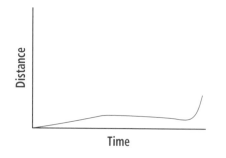

2. a. Describe the relationship shown in this graph. Make up a story about the cyclist that could be modeled by the graph.

  b. Over what interval in the graph is the cyclist traveling the fastest? Explain how you know.

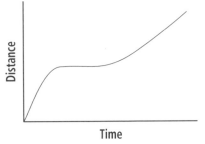

3. a. Describe the relationship shown in this graph. Make up a story about the cyclist that could be modeled by the graph.

  b. Speculate about what will happen next, and explain how the graph would change.

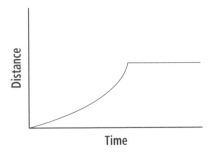

Writing Stories to Match Graphs

In the last problem, you matched graphs with real-life events described in a story, and you sketched graphs to match written descriptions. In the follow-up, you made up stories about what given graphs might represent. In this problem, you will continue to work with graphs of real-life events.

A car and a bus are traveling in the same direction on the same road. The bus has just stopped to pick up a passenger, and the car has stopped directly behind it. In this problem, you will look at several graphs that show what might happen next, and you will make up stories that the graphs could represent.

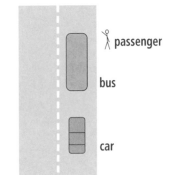

Graphs A–F show six possibilities for what might happen at some point after the passenger boards the bus. Each graph shows the relationship between time and distance from the bus station for the bus and the car. For each graph, make up a story about the bus and the car that matches the information in the graph.

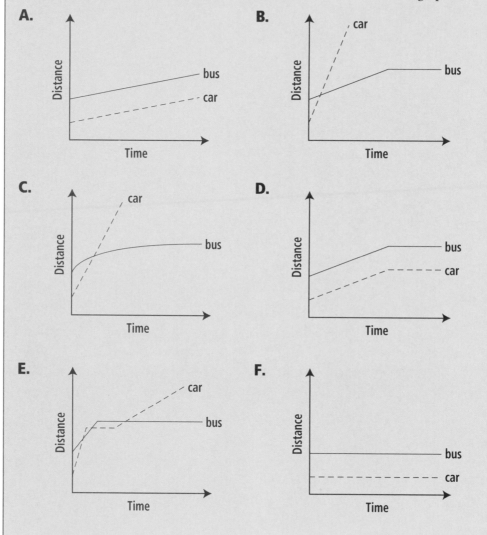

A.

Distance / Time — bus, car

B.

Distance / Time — car, bus

C.

Distance / Time — car, bus

D.

Distance / Time — bus, car

E.

Distance / Time — car, bus

F.

Distance / Time — bus, car

Which graph do you think best represents what might really happen? If none of the graphs seem correct, sketch a graph of what you think is likely to happen.

■ **Problem 4.2 Follow-Up**

1. In which graph does the car reach its greatest speed? Explain how you know.

2. In which graph does the bus reach its greatest speed? Explain.

4.3 Exploring Graphs

A graphing calculator is a useful tool for exploring graphs of interesting equations. Making graphs of equations by hand can take a lot of time. With a graphing calculator, you can enter an equation and see its graph almost instantly.

Problem 4.3

A. Use your graphing calculator to graph each equation below. Adjust the window settings until you think you have a good view of the graph. Copy the graph onto your paper.

1. $y = \frac{1}{x}$　　　　　　　　**2.** $y = (x - 1)(5 - x)$

3. $y = 2.7x$　　　　　　　　**4.** $y = 2^x$

B. Choose two of the graphs from part a, and make up stories that could be modeled by them. For each graph you choose, be sure to tell which variable in your story is on the horizontal axis and which variable is on the vertical axis.

■ Problem 4.3 Follow-Up

1. a. Do you think every real-life situation can be modeled by a graph?

b. Do you think every graph can be represented by an equation?

2. Look back at the graph models you have made in this unit. For each graph you made in part A of Problem 4.3, find a graph model from your earlier work with a similar shape. Make a sketch of the graph model you find, and give its equation if you know it.

As you work on these ACE questions, use your calculator whenever you need it.

Applications

In 1–3, do parts a and b.

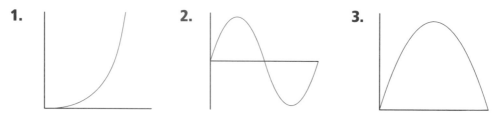

1. **2.** **3.**

a. Describe a situation you think could be modeled by the graph. Be sure to tell which variable each axis represents.

b. Explain why the graph is a good model for the situation you described.

4. The table below shows Jamie's height from age 5 to age 18.

Age	5	6	7	8	9	10	11
Height (cm)	110	116	121	127	132	137	143

Age	12	13	14	15	16	17	18
Height (cm)	149	157	165	173	176	178	180

a. Graph the (age, height) data. Draw a straight line or curve to model the data.

b. What was Jamie's average rate of growth from age 5 to age 12? If this rate of growth had continued, how tall would Jamie have been at age 18? How tall would he have been at age 25?

c. During what time period did Jamie grow the fastest? How can you see this in the graph?

5. The drawings below show the progress of a tour bus over a 35-second period.

a. Describe what is happening in the drawings.

b. Sketch a graph of the relationship between time and distance from the starting point for this series of drawings.

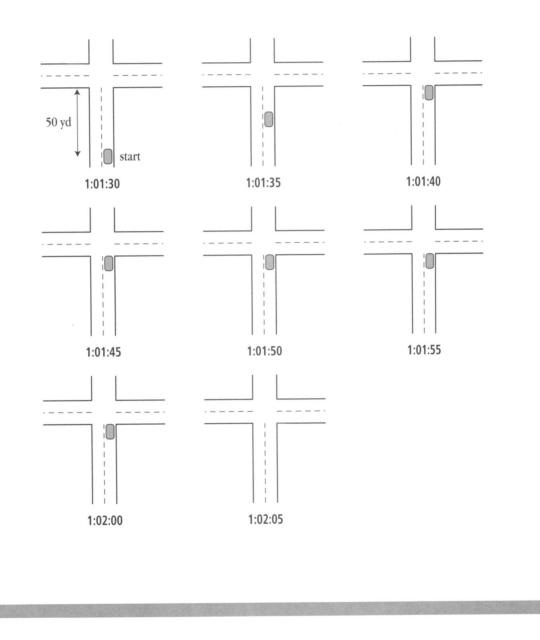

Connections

6. The graph below shows the progress of the Orient Express on a trip from London to Paris.

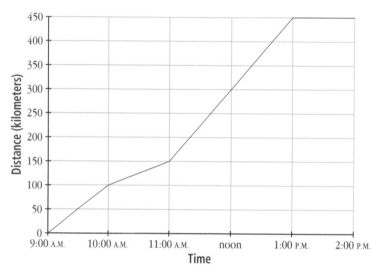

a. Describe in words what the graph shows.

b. Write an equation for the part of the graph between 9 A.M. and 10 A.M.

c. Write an equation for the part of the graph between 10 A.M. and 11 A.M.

d. Write an equation for the part of the graph between 11 A.M. and 1 P.M.

Big Ben in London

The Eiffel Tower in Paris

7. In your previous math work, you investigated the relationships among the radius, height, base area, and volume of a cylinder. You found that the volume of a cylinder is equal to its base area multiplied by its height.

a. Suppose you are in charge of designing a cylindrical can to hold 250 ml of juice. Investigate some possible (base area, height) combinations for the can. Try radii of 2.5 cm, 3 cm, 3.5 cm, and any other measurements you think are reasonable. Record your findings in a table like the one below. (Recall that 1 ml = 1 cm^3.)

Radius (cm)	Base area (cm²)	Height (cm)	Volume (ml)
			250

b. Make a graph of your (base area, height) data.

c. Draw a straight line or curve to model the data. What other situations in this unit have similar graph models?

d. Write an equation that fits your graph model.

e. Which (base area, height) combination would you choose for the can? Give reasons for your answer.

Did you know?

Designing a can is a complex process. Manufacturers must consider the cost of the sterilization process, the strength of the can, the amount of waste metal left after cutting the base and side, and the packing and shipping space the cans require. Here are three important relationships they must think about:

- There is an *inverse relationship* between the amount of material used and the sterilization time. In other words, for a given volume, cans with less surface area take more time to sterilize than cans with greater surface area.

- Cans with wider bases must be made of thicker metal so they will be strong enough to withstand the pressures of the sterilization process.

- Cans with smaller diameters waste less space in packing and shipping.

8. Terri is designing a can. She has collected the following data about diameters and sterilization times.

Diameter (cm)	1	1.2	1.8	2.4	2.8	3	3.2	3.5	3.8
Time (seconds)	6.2	5.2	3.5	2.6	2.2	2.1	1.9	1.8	1.6

a. Make a graph of the data by hand or by using a graphing calculator. Draw a straight line or curve that models the data. Which other problems in this unit have similar graph models?

b. Find an equation that fits your graph model. If you are using a graphing calculator, enter your equation and check the fit.

c. Use your graph or equation model to predict the sterilization times for a can with a diameter of 1.5 centimeters and a can with a diameter of 4.5 centimeters. Explain how you made your predictions.

Extensions

9. The diagrams below document the growth of a pea root at regular intervals.
   Use these diagrams to answer parts a and b on the next page.

Diagram 1

Diagram 2

Diagram 3

Diagram 4

Diagram 5

Diagram 6

Diagram 7

Diagram 8

Diagram 9

a. Make a graph of the relationship between the diagram number and the length of the pea root.

b. Draw a straight line or a curve that models the data. Is your graph model linear? Explain.

10. Some of the equations below have appeared in problems you worked on in this unit. The other equations have graphs with the same general shapes as graphs you have investigated.

a. Without graphing the equations, organize them into three groups according to the shapes you predict for their graphs.

b. Use your graphing calculator to graph the equations. (Suggested window settings are given for each equation.) Make any adjustments necessary in your groupings.

i. $y = 8.7x$ (suggested window: $x = 0$ to 10, $y = 0$ to 100)

ii. $y = {}^-40x + 480$ (suggested window: $x = 0$ to 14, $y = {}^-50$ to 500)

iii. $y = \frac{120}{x}$ (suggested window: $x = 0$ to 100, $y = 0$ to 100)

iv. $y = \frac{8.7}{x}$ (suggested window: $x = 0$ to 20, $y = 0$ to 20)

v. $y = 100(1.08^x)$ (suggested window: $x = 0$ to 20, $y = 0$ to 500)

vi. $y = \frac{40}{x}$ (suggested window: $x = 0$ to 50, $y = 0$ to 50)

vii. $y = 1.08x$ (suggested window: $x = 0$ to 10, $y = 0$ to 10)

viii. $y = 100(0.5^x)$ (suggested window: $x = 0$ to 10, $y = 0$ to 100)

ix. $y = 100 - 1.08x$ (suggested window: $x = 0$ to 100, $y = 0$ to 100)

x. $y = 2^x$ (suggested window: $x = 0$ to 10, $y = 0$ to 100)

xi. $y = 2x$ (suggested window: $x = 0$ to 10, $y = 0$ to 10)

xii. $y = \frac{2}{x}$ (suggested window: $x = 0$ to 10, $y = 0$ to 10)

c. For each group of equations, make up a different equation that you think belongs in the group. Use your calculator to check your equations.

Mathematical Reflections

In this investigation and throughout this unit, you looked at relationships associated with real-life situations. You found that many of these relationships can be represented by graph models and equation models. These questions will help you summarize what you have learned:

1 Look back at the graphs you have made in this unit. Find several graphs that show relationships in which *y* increases as *x* increases. Try to find graphs with as many different shapes as you can. Sketch each graph, and describe it in words.

2 Look back at the graphs you have made in this unit. Find several graphs that show relationships in which *y* decreases as *x* increases. Describe each graph in words.

3 Look back at the graphs you have made in this unit. Find several graphs that show relationships in which *y* both increases and decreases as *x* increases. Describe each graph in words.

Think about your answers to these questions, discuss your ideas with other students and your teacher, and then write a summary of your findings in your journal.

Glossary

equation model An equation that describes the relationship between two variables. In this unit, you fit graph models to data points, and then, when possible, you used your graph models to find equation models. An equation model allows you to make predictions about values between and beyond the values in a set of data.

fulcrum The balance point of a teeter-totter or balance. In the teeter-totter experiment, you measured distances from the fulcrum to the weights on both sides. You found that for a teeter-totter to balance, the product of the weight and distance on one side of the fulcrum must equal the product of the weight and distance on the other side of the fulcrum.

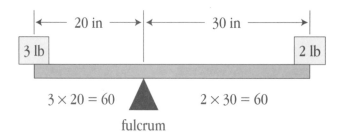

fulcrum

graph model A straight line or a curve that represents a mathematical relationship. If the data you plot show a trend, you can draw a graph model that fits the pattern of change in the data. A graph model allows you to make predictions about values between and beyond the values in a set of data.

inverse relationship A nonlinear relationship in which the product of two variables is constant. In an inverse relationship, the values of one variable decrease as the values of the other variable increase. In the bridge-length experiment, you found an inverse relationship between length and breaking weight. In the teeter-totter experiment, you found an inverse relationship between distance and weight.

linear relationship A relationship in which there is a constant rate of change between two variables. A linear relationship can be represented by a straight-line graph and by an equation of the form $y = mx + b$. In the equation, m is the slope of the line, and b is the y-intercept.

mathematical model A mathematical representation, such as a graph or an equation, of the relationship in a set of data.

relationship An association between two variables. A relationship can be represented in a graph, in a table, or with an equation.

Index